0

ISBN: 978-0-557-36133-5

The Universe is Nuts

Unfortunately That Might Include Me

Original random thoughts - from the pen of G. Donald Mitnik

Afar they've sailed with wings on high
From other realms and times,
to take the Earth from our grasp
and steal our very minds.

- G. Donald Mitnik

As the contents of this book will attest, there is an unseen force working overtime to infiltrate and numb all our brains to the point of incoherence. I suspect "they" have a diabolical plan to steal all the Earth's limburger cheese to take back to their planet and sell as fuel for their personal propellant machines. I would like to present the following as undeniable proof of mankind's slow but sure plunge into the abyss by way of outside intervention. Be warned: This book alone will, as "they" have planned, drive you to the brink of insanity. Please proceed with caution.

Introduction

I've always wanted to write down some of the bizarre thoughts that pop into my head from time to time, but being a bit of a procrastinator, I never seemed to get around to it. So, one day I decided – if I could just write a few thoughts per day, I would eventually have enough stuff for a book. Surprisingly, it wasn't long before I had 1,001 of them starring up at me. Some of what you read here is seriously profound. Other things, although thought provoking, are somewhat inane. This book is not meant to be philosophical or contain words to live by. It is simply a series of one or two sentence questions and statements about humanity and the universe. They are all absolutely original creations, never put in print before now. This is not to say they were not thought of before they flashed through my meager brain. I've always had the notion that if I'm thinking of something, somewhere in the world, someone else is having exactly the same thought - more than likely it's some guy in an alley with a wine bottle in his hand. Other bits and pieces have been spinning in my head for decades and like energy bouncing around in the Sun's core have finally made it to the surface. The truth is; I really don't know how or why these thoughts occur to me. My guess is; they are being transmitted by extraterrestrials bent on causing all our brains to meltdown. In any case, they have manifested themselves in this dysfunctional array that you now see before you. My only hope is; you come away unscathed.

I know a vast amount of idiocy is on the internet these days and most of it has made the rounds many times over. I can assure you, none of what is written here has been siphoned off the internet. For better or worse it has all come to me somewhat like a medium receives messages from the "other side". All these mixed up questions and answers to questions nobody asked have arrived randomly without continuity or relevancy so that is exactly how I decided to write this book. All these tidbits appear in the order they bounced out of my head. No index is needed or wanted. If there is an item you wish to remember simply write its number down for future reference. My only purpose for writing this book is to cause you the same brain numbing wonder I experience daily and if you find any of the contents offensive or demeaning, all I can say is; "Damn extraterrestrials".

Now pick a page and start reading. Since there is no organization in the writing, you're free to wander through these pages in any order you like. My advice is to have fun with these outbursts of energy and above all take heed knowing that, in all probability, they arrived from another realm and time.

No doubt I'll continue to be the recipient of a variety of stealthily generated questions and revelations that prove the downfall of human thought is at hand and I'm sure you'll be happy to know I will continue to put these forth coming creations down on paper as they occur. So, look for Volume II of these never-ending bits of insight sometime in the future. In the meantime I'll try to stay away from wine bottles, the Sun and any uninvited extraterrestrials.

The Universe is Nuts
Unfortunately That Might Include Me

Original random thoughts - from the pen of G. Donald Mitnik

Common sense will now be asked to leave... Logic can stay but please refrain from scratching your head. For those of you who are left, in numbered (although chaotic) order of conception, I present the following array of extra-planetary froth into which I invite you to soak your brain.

---

1. The universe sings through us. We just call it music.

2. Men reaching for Mars are like fish reaching for Bakersfield.

3. Interesting people are not always interesting.

4. Without language, all thought is abstract.

5. If the stars shown for only one night every 10,000 years, would it be enthralling or terrifying?

6. If one speaks the truth and none believe it, is the truth any less diminished?

7. Without humanity, the universe has no benevolence or consciousness.

8. If you imagine it - it can exist. If it exists you can't always imagine it.

9. Walking on water is hard. Walking on Earth is unbelievable.

10. The most astounding thing in the entire universe is you.

11. What would the scent of a rose smell like if there were no roses?

12. Education is better than wealth. You can spend it many times over without it ever leaving you.

13. If you expect more, you'll always receive less. If you expect less you'll always receive more.

14. Never hold a nail for someone else to hammer.

15. A thousand years from now will anyone know who Einstein was?

16. The next big thing is downsizing.

17. If you need a friend, draw a face on your thumbnail.

18. If all the world is a stage, where's the audience?

19. That which is beyond comprehension - yet exists, will either be worshiped or ridiculed.

20. There is a 100% chance your perception of reality is wrong.

21. Why is "forever" a longer time than "never"?

22. Professional wrestling is real - because it makes money.

23. If a company called "Mother Goose Cook Books", cooks its books, will their goose be cooked?

24. Do they really need Viagra in China?

25. People are spending a fortune trying to find the best way to save money.

26. All things change - except change itself.

27. What were you doing five years before you were born?

28. If humans never saw things in color, would humans know color existed?

29. It is not "what would Jesus do?" Today, it's what would we do with Jesus?

30. Why is the world round and the galaxy flat?

31. In an infinite number of universes with an infinite number of possibilities, we've all had sex with Pamela Anderson.

32. It's a fact - spirits never have to go to the bathroom, but do they ever have to do their laundry?

33. You can exist out of your body as easily as you can exist out of your mind.

34. Who on God's design staff is responsible for the external meanderings of human ears?

35. The atomic structure of the cosmos is interesting – but, to most people, football is far more important.

36. If something can be nothing and nothing can be something, is anything really anything?

37. What makes us think life is the highest manifestation we should be looking for in the universe?

38. If you agree entirely with someone else – it's time to rethink your position.

39. A politician is the only salesman that can get away with selling you nothing.

40. Most knowledge is unknowable.

41. If space/time is a fabric, what could be hiding in the wrinkles?

42. The illusion of time is real – time itself is not.

43. Keeping a stiff upper lip makes eating soup difficult.

44. Do viruses get viruses?

45. The person with the greatest intellect in the world today is probably picking rice somewhere in Asia and will be doing so for the rest of their life.

46. Why do we always think in our own voices?

47. Investing in the stock market is like betting on a horse race without the benefit of the race.

48. If your kids are building a snowman, never give them more than one carrot to use.

49. Mankind's advances in science and technology are far outweighed by his ignorance in every other facet of living.

50. Whether you're born poor and die rich or born rich and die poor, the end result is just the same.

51. Eventually, there will come a time when a loaf of bread will cost as much as a house does today.

52. Do you ever wonder if people with blue eyes see the world with a slightly blue tinge or people with brown eyes see it with a brown tinge?

53. Do people really buy products just because they saw them on TV?

54. Spilling hydrocarbons into the atmosphere is like spitting into the gravy. It probably won't hurt you and you won't notice, but it's still spit in the gravy.

55. Avoiding the fact that an asteroid could hit the Earth is making a molehill out of a mountain.

56. Shouldn't the color for the environmental movement be clear?

57. To avoid the consequences of 12-21-2012, why don't we just skip from 12-20-2012 directly to 12-22-2012? (depending on today's date, maybe we did.)

58. People who like guns also tend to be drawn to knifes, brass knuckles, crossbows, slingshots and nuclear devices.

59. Some people have lives that are so good – they must manufacture grief.

60. Everyone has stolen something. If you believe you're an exception, you're just not thinking hard enough.

61. While allied troops were bombing Berlin, there were a substantial number of residents preparing to go to the opera.

62. Without war, the word "peace" wouldn't mean as much.

63. Is Christmas really the best time for charities to ask for money?

64. Mediocrity will always rise to the top.

65. Attention NASCAR: Segway racing would make a fortune.

66. If you reach for the stars you'll get a crick in your neck.

67. You will substantially increase your chances of winning a lottery if you buy a ticket.

68. Good health is an investment that you are continually taking profits from.

69. If you pack a one pound bag with five pounds of (anything) - wear a raincoat.

70. The top rung of a ladder is also the most dangerous.

71. If the grass is always greener on the other side of the fence, it's time to spread more manure around.

72. There is some value in being a little bit crazy.

72. Nothing in life beats a good pastrami sandwich. (Well, almost nothing.)

73. Why is the back burner less important than the front burner?

74. A thesaurus proves the human need for redundancy.

75. How far can the fabric of space be stretched before it rips?

76. If we flipped the universe over, would south then be up and north down?

77. Wouldn't you think we'd know when we kid ourselves?

78. Dogs possess all the knowledge of the universe. They just can't express it.

79. Cats think they're smarter than dogs.

80. As Wal-Mart goes, so goes China.

81. The greatest facade the government ever perpetuated on the people is the government itself.

82. If someone could be transported from 1942 to the present time, it would be hard for them to tell who won WW-II.

83. There's the possibility that one hundred years from now, we'll all be driving Iranian cars.

84. At what point in time and scale does a mini black hole become a maxi black hole?

85. In the game of life, we could be running toward the wrong goal.

86. Somewhere there's a more advanced civilization than ours that has never seen what lies beyond their cloud shrouded planet.

87. When a writer spares a reader the tedium of a cliché, it's a win-win situation.

88. There is no explanation for gravity, yet here it is.

89. If you hang around long enough, people will start listening to you – only if it is just for the laughs.

90. Tradition is perpetuated by those who sell cards, candy, flowers, flags, fireworks and turkeys.

91. If you want to know where the big bang happened, look deep into the period at the end of this sentence.

92. The future evolution of basketball is evident. It's time to either raise the basket or lower the floor.

93. If a religion that promotes killing, dominated the world, would the words "good" and "evil" reverse their meanings?

94. Man is the only animal that can fly without flapping anything.

95. Many people wake up at the crack of doom.

96. If you were born into the Wallenda family, would you walk on a high wire, 500 feet in the air without a net just because it was expected of you?

97. What is man that he gets wine-full once again?

98. Discovering the unknown can change the meaning of all that is known.

99. Language, art and imagination is what puts man above all other beasts. It's the last of these that also gets him in the most trouble.

100. The invention of the wheel was not that great. The real breakthrough came sometime later with the invention of the axle.

101. The news media exists to make money, not to keep you informed.

102. Magicians prove we can all be manipulated, fooled and misled.

103. To survive, American car companies should just put the best of their concept cars into production.

104 Liberals in congress think conservatives are crass. Conservatives think the same of liberals. In reality, the "cr" in crass should appropriately be dropped.

105. The three legged stool is the result of compromise.

106. Over time, a nuclear war is inevitable.

107. What did we do to waste time before the internet?

108. To make a long story short is easy. To make a short story long is preposterous.

109. It's hard to believe Christmas is only one day.

110. Mankind is the means by which the universe contemplates itself.

111. If Los Angeles ever gets a professional football team, they should be called the At-LA-sts.

112. How did the universe know we were coming?

123. What changes would you make if you were given the opportunity to design a new human?

124. Why do people think puns need to be pardoned?

125. If you wear tennis shoes while playing basketball, will net balls be played over?

126. The only similarity between astronomy and astrology is: They both involve the sky.

127. Disregarding religion, the greatest narrative in history has probably been long forgotten.

128. People with tinnitus experience life at a higher frequency.

129. What compels people with no opinion to vote?

130. Does "the point of no return" always have to be half way?

131. If an envelope is made in Manila, but its color is white, is it still a Manila envelope?

132. If the government decided to print counterfeit money, would it really be counterfeit?

133. The great pyramid of Giza in Egypt was actually a giant pomegranate processing plant.

134. There is no known cure for immortality.

135. If the "sky's the limit," the goal is set much too low.

136. Why are communists – Red?

137. Why are depressed people – Blue?

138. Why are environmentalists – Green?

139. Right now, there's a small company starting out, looking for investors. A $1,000 investment in that company today will be worth millions in ten years. We just don't know what company it is.

140. Galaxies are giant dust bunnies.

141. We all have a mind of our own. Geniuses have theirs - plus someone else's.

142. Why isn't "Blackbottom" a professional sport?

143. For millions of years, the ultimate weapon was a rock. In some parts of the world it still is.

144. In the 1940s, shoe stores would routinely x-ray your feet.

145. Mankind has a long past, but the future is likely trillions of times longer - making the past insignificant.

146. Once upon a time, plumbers wore ties to work.

147. When upset, dogs growl, cats hiss, monkeys hoot and turkeys gobble. Humans do all of these when they're feeling good.

148. If you were to travel to the nearest star and look up in the night sky, you would see no discernable difference in the position of the stars.

149. Any conspiracy theory has, at least, an outside chance of being true.

150. Shouldn't the first class on the first day of school be called disorientation?

151. If we lived on Mars instead of Earth would we call ourselves Marslings and wonder if Earthtians existed?

152. If a home run is hit out of the park and lands in a pile of garbage, shouldn't it be considered a foul ball?

153. Without man's continual extermination efforts, Cockroaches would soon overrun the entire Earth.

154. Chem-trails are a cover story.

155. Is there anybody today with the last name - Hitler?

156. When you dream, the reasoning part of your brain is turned off.

157. Science is continually hobbled by the paradigms of known physics.

158. Anyone willing to die for a cause is psychotic.

159. What compels people to write letters to the editor, thinking they're going to change things?

160. There are times when a novel would be better if it were written backwards.

161. A "Guest of Honor" implies the other guests are dishonorable.

162. In theory, you can bet there are theories of theoretical theocracies.

163. Ants keep mites as pets.

164. Why is the time to sit back and enjoy the fruits of your labor, always tomorrow?

165. You would think a human would be more efficient if built on the monocoque principal instead of a skeletal frame.

166. What does a raving maniac usually rave about?

167. When you back your car up, do you become the back seat driver?

168. If you need more sleep, watch more television.

169. The fear of obscurity is only eclipsed by the fear of poverty.

170. If the speed of light was cut in half or doubled would there be any difference in our daily lives?

171. The lobbyist influence in government is the highest form of corruption.

172. Weapons bring out the most inventive ideas in humans.

172. If scientists searched right the first time, they wouldn't have to research.

173. So far, the strangest discovery ever made by man is the ground he's standing on.

174. Why are wooden pencils still around while rock candy on a string is gone?

175. Give a man enough soap on a rope and he'll hang himself.

176. If an athlete praises Jesus for a win, who will he hold responsible for a loss?

177. Shouldn't the first guy to try a parachute be famous?

178. There are persons who prefer the isolation of prison compared to what they might face in the outside world.

179. Never race anybody nicknamed – Bullet.

180. The most responsible job in the country belongs to the American Consumer.

181. Didn't get what you wanted for Christmas? Next time, instead of milk and cookies, leave Santa a shot of Seagram's VO.

182. There is no such thing as a moment in time.

183. The more annoying a TV commercial is, the more times it is shown.

184. Why isn't the dollar sign a capital D with a vertical line through it?

185. If you're outside in a major city, chances are there's a camera on you. If you're in a store or public building, it's for sure.

186. History isn't what it used to be.

187. The highest compliment any artist can attain is when someone steals their work.

188. Was Crusader Rabbit an affront to Islam?

189. In a parking lot of 500 cars, why do you and the person parked next to you always leave at the same time?

190. Portugal was once a world power.

191. If there is a door between two rooms and you go through it, are you going in or out?

192. Wouldn't it be more appropriate to call cowboys, horseboys?

193. Dark matter is not necessarily dark.

194. Why is showing your bare posterior to someone called mooning when a more descriptive term would be catcher's mitting?

195. Is it better to have charisma or chutzpah?

196. If you're not in a pickle and not in a beef and not in a stew, you're probably in the pink.

197. A good way to measure a person's fame is by the size of their entourage.

198. Is "catch as catch can" really a type of wrestling?

199. How does anyone know if a tree pines?

200. To receive exceptional rewards, you must take exceptional risks.

201. If humans had as much hair as gorillas, would we still wear clothes?

202. The most destructive thing in the universe is the human mind.

203. Opportunity knocks on random doors.

204. Ideas are concepts that start as a spark in the mind. Some catch fire, others don't.

205. A person's plans seldom end as they began.

206. Very few corporations last more than 100 years.

207. The most noble of us all are those who are gainfully employed doing the most mundane jobs.

208. If you bought a new car in the 1950s, optional equipment would include turn signals and a radio.

209. Eating Indian fry bread should be a felony.

210. Why are the indicators on a clock called hands?

211. The fact that viruses and disease abound is proof that the universe does not exist entirely for man.

212. Why are streets named "Broadway" no wider than other streets?

213. Shouldn't insanity actually be called: outsanity?

214. We experience what is called reality for a fleeting instant while eternity is forever. Which one do you think is the illusion?

215. Never play chess with a guy who smokes a pipe.

216. The more you accomplish, the more you are expected to accomplish.

217. Do we ever stop wanting the newest toy?

218. Always look forward to a challenge – especially when you know it can't be met.

219. Those under age 65 might not know: Medicare recipients pay premiums for their insurance and taxes on their Social Security – which they bought and paid for while they worked. (...and some call it an entitlement?)

220. Progress depends on your point of view.

221. Global warming, either man-induced or by nature, will (counter intuitively) produce a new ice age.

222. Except for being a substitute for finger thumping, bubble gum has no redeeming value.

223. Never fix a rocket engine while it's running.

224. The difference between a terrorist and a patriot lies in the minds of those whom sponsor him.

225. It's healthy for you to feel a little guilty when you step on a bug. For the bug – it's not so healthy.

226. Did Neanderthals speak French?

227. The problem: The good think they are good. The evil think they are good. The answer: Education.

228. Intellect seeks its own level.

229. Every minute you're on an airplane, you're trusting a stranger with your life.

230. When someone wants a calendar date to be unknown, they'll express it in roman numerals.

231. There had to be some new car designs that were meant to be light hearted jokes among designers. Apparently, management was not in on the joke.

232. Bipedalism is a balancing act.

233. Intelligence should never be confused with wisdom.

234. The longer you wait to clean your glasses, the clearer the world will look when you finally do.

235. In a recession, the "Peter Principle" somehow peters out.

236. The population explosion seems to have a longer fuse than first thought.

237. All pantomime is the same when preformed on the radio.

238. A collage degree is never a substitute for the ability to articulate ideas.

239. Everyone who is either a skeptic or advocate of climate change will be dead by the time the facts become evident.

240. If you have a choice, always take the scenic route.

241. Integrity is not a goal, but a way to obtain a goal.

242. When a clown isn't funny he'll blame his character. When he is funny he'll take the credit himself.

243. A forgetful person will always be full of new ideas.

244. Cynicism is the result of seeking perfection.

245. A blank canvas is never more intimidating than a canvas that is painted badly.

246. Anyone born with a silver spoon in their mouth should see a surgeon immediately.

247. If you argue with someone who happens to be right, stand your ground. Eventually someone will support you.

248. There are no two experts who agree entirely on anything they are expert in.

249. Embellishing the truth is not quite the same as lying. It's usually more unbelievable.

250. Where matter came from will always be a mystery. What is known: Matter cannot be the result of non-matter.

251. Scraping the bottom of the barrel will wear out the barrel – and your scraper.

252. Was the U.S. cost for WW-II ever an issue?

253. If the glass is either half full or half empty, someone else should do the pouring.

254. Without humans, what would be the reason for cocoa's existence?

255. Magic is a form of brainwashing.

256. Fate is predetermined – but only by seconds.

257. When a musician makes a mistake playing drums, will there be repercussions?

258. Do you suppose anyone ever contemplated "buzzing" the President while shaking his hand?

259. Tea leaves will predict your next choice of tea.

260. If it's truly better to give than receive, those who were in the stock market during the first decade of the 21st century must feel warm and fuzzy all over.

261. When you're in a supermarket it's ok to compare apples to oranges.

262. Would Magellan have trouble navigating the web?

263. No organization, except organized crime, operates with more complete disregard for the will of the people than the U.S. congress.

264. In an upscale restaurant, a person's social status can be determined by how far they are seated from the rest rooms.

265. It wasn't that long ago when mercurochrome, a substance containing mercury, was used as first aid on open cuts.

266. From two tin cans connected by a string, to Skype. Isn't alien technology wonderful?

267. Demolition Derby and Figure 8 Racing could only have originated in the U.S. – the only place where there's an audience for it.

268. When disappointments are called successes, how will a true success ever be known?

269. Fame is a lot like owning a pet. It must be continually nurtured. At times it's a nuisance. If not treated right it can bite you and it usually dies before you do.

270. The most gifted children can become the most mundane adults …and vice versa.

271. When Brigham Young picked Utah as the Promised Land, there was an outside chance he had a world land speed record attempt in the back of his mind.

272. A bird in the hand is worth keeping a tissue nearby.

273. It's no *secret* how many times the same idea can be rephrased over and over again. What is also no *secret* is the number of people who want to hear a *secret* because they were told it's a *secret* even when *the secret* is not really a *secret*.

274. Costs should never stand in the way of a needed prison.

275. The U.S. has more prisoners per capita because in outer countries they'll either just kill you or chop off your hands for committing a crime.

276. There is a free lunch. Ask any pigeon.

277. We only see groupings of atoms. We can't see the realities that might lie in the vast space between them.

278. The most important meal of the day is the one you miss.

279. If you're ever abducted by space aliens, by all means steal something to bring back with you.

280. A clock doesn't measure time. It measures other clocks.

281. All myths are founded in truth.

282. The key to solving a mystery is to fully understand what surrounds the mystery.

283. If it wasn't for the dust in the galaxy, the sky at night would be as bright as day.

284. Would envious environmentalists be two shades of green?

285. If terrorists believe dying is life's most noble accomplishment, why do they seek death for those they perceive as their enemies?

286. Is there a difference between Nutty Putty and Play Doh?

287. All history is a prelude to more history.

288. If a "W" is called a double U, why isn't an "M" called a double N?

289. The greatest boon to the progress of mankind was the invention of the bank.

290. Saddam Hussein and Adolf Hitler were bad guys who didn't know how to play the game. Today, there are many who do know how.

291. Minimum wage has the potential to be a lofty goal.

292. The whole point of why we exist might be to figure out the whole point of why we exist.

293. Why is the oldest living thing on Earth, a plant?

294. From the Lone Ranger Atomic Bomb Ring to World of Warcraft in just 60 years. We sure know how to keep kids entertained.

295. People who pack a gun will subconsciously look for a reason that will justify carrying it.

296. It is perplexing that no state in the U.S. endows its official state seal with a seal.

297. No matter where a farmer shops, It's a Farmer's Market.

298. In the world of the deaf, sound is unknown and unneeded.

299. A child prodigy cannot accurately tell you how their abilities were acquired.

300. Antiques are our only tangible link to what we remember.

301. All things that are thought of as true today - might not be true tomorrow.

302. Like catching the flu, anyone can become a sociopath for a time.

303. Language is the paint, speech is the brush, poetry is the painting.

304. There has to be occasions when a race car driver sneezes while wearing his full face helmet.

305. Do you think dodging our space junk is a problem for extraterrestrials?

306. In a nuclear scenario, MAD (mutually assured destruction) will not work if self-destruction is one side's desired goal.

307. Because of funding sources, Scientists do not truly have freedom of speech.

308. The appearance of the trophy doesn't always match the accomplishment of the feat.

309. If you get killed by a car while jogging, would the jogging still have been beneficial?

310. Going for a prostate exam can be uplifting.

311. If you want to see more of someone, borrow money from them. If you want to see less of someone, loan them money.

312. Why is a handkerchief poking out of a suit coat pocket considered stylish?

313. Has there ever been a survey to determine how many people lie when taking surveys?

314. Never stand in the way of plate tectonics.

315. Wasting calorie intake on something bland is poor dieting.

316. The future is not determined by those that will be here in the future. It is determined by us.

317. When aiming for a goal, always allow for windage.

318. A park in the middle of a city is what people need in the middle of their minds.

319. Value is not determined by actual scarcity, but by what individuals think has a potential for scarcity.

320. The wheels of progress sometime go off- roading.

321. There are times when we all deserve a pie in the face.

322. The high road requires more power to attain.

323. Be careful dropping hints, they can bounce back at you.

324. There is no difference between night and day. The difference lies in your perspective.

325. Patience is a virtue, but only if you have time to spare.

326. From singularities to entanglement, there's no other rational alternative: The universe is nuts.

327. Socrates said he didn't know anything, but he certainly knew how to get published.

328. Extremist Islamists are like the Nazis – empirical, racist elitists and, over time, their movement is doomed to the same fate.

329. One of the most turbulent and violent formations in the universe is only 1800 miles away – the Earth's core.

330. At the atomic level, there is more space within a person's head than matter.

331. Ask any prison guard, the amount of stuff a person can hide in their rectum is ass-tounding.

332. From moment to moment your location in space is continually moving - never to return to where you once were.

333. Torque cannot be explained without a twist ending.

334. Logic is the end result of intellect. Wisdom is the end result of logic.

335. If you "Follow the money" - Chances are it will lead from your wallet to the government.

336. Fake collage degrees are worth exactly their price.

337. There is one word in the English language that rhymes with orange. It is a secret and is known only to a 98 year old English professor at Oxford.

338. A hermit is someone who knows the true value of civilization.

339. Considering the daily volume of mail that is processed, the Post Office is actually an extremely competent organization.

340. The last surviving human on Earth will be a politician.

341. Everyone over age 70 should be made King of somewhere and addressed as: Your Magnificence.

342. If an efficiency expert is not efficient, is he then an inefficiency expert?

343. Perception is everything reality is unclear about.

344. Con-men are such polite bastards.

345. The Tower of Babel was hard to describe in words.

346. The current state of affairs among our elected officials seems to be rising.

347. Without communication satellites, all live transfer of information beyond your voice would cease.

348. Your accomplishments define who you are. Your dreams define who you can become.

349. A lot of people make a six figure income, if you count the cents.

350. Wouldn't we all sleep easier if an alarm clock was called a gently wake you up clock?

351. The fall of Rome began with the indifference of the people.

352. Could a universe without matter still be called a universe?

353. Having to mop up a messy spill goes beyond the pail.

354. Economically speaking, inflation is bad for expansion.

355. Genius can blossom from the most fertile soil, or the most desolate.

356. Television really didn't become popular until the TV program was invented.

357. As hard as we try to control the environment, The Earth seems to do what it wants.

358. Ignorant people are so imperceptive they don't know they're ignorant.

359. Political correctness is not always political and seldom correct.

360. Isn't it strange that millions of reasonable people believe the impossible in order to gain a theoretical ticket to an ethereal eternal life?

361. The laws of gravity can weigh on you.

362. Grey comes in a vast array of shades that we can't always see.

363. If you jump to conclusions you should at least know how far off target you could land.

364. A wish is a prayer that you don't think is serious enough for God's ears.

365. Why does the mail move slower when you're expecting a check?

366. Why do people travel for hours on a highway at 75 miles per hour in order to stand in line for an amusement park ride that lasts 2 minutes at 60 miles per hour?

367. Despotism, war and terrorism all have entertaining qualities.

368. Irresponsibility takes years of nurturing to get right.

369. Making money can sometimes cost a fortune.

370. A slow news day is good – unless your business depends on TV ratings.

371. Common sense is not so common.

372. The problem with credit cards is: They don't seem to be a problem.

373. There's no present like the time.

374. Why do Levis cost twice as much as Wranglers?

375. Electricians and plumbers have connections.

376. How long should a take last before it becomes a double-take?

377. When you consider we spend 30 years of our lives sleeping, think of all the stuff we must have missed.

378. A Peacock is the result of one-upmanship.

379. Ten years can sometimes seem like a whole decade.

380. The Town Crier of old should have been called the Town Yeller. Anybody trying to sleep would have been the Town Crier.

381. After trying A, B, C, D, E, F and G, on the eighth try, finally - Preparation H was born.

382. If a left handed person is facing east, would they still be a southpaw?

383. The ideal example of the phrase - "Easy come, easy go" - is when you are written out of a will.

384. A lie detector detects lies. What it can't detect is the truth.

385. The power of positive thinking, or positively thinking of power, should produce the same result.

386. Known secret societies are a contradiction of terms.

387. The best way to meet your neighbors is to have a garage sale.

388. The need for three meals a day is a creation of the food and restaurant industry.

389. The wave of the future might be a tsunami.

390. Propaganda is the truth seen from very far away.

391. The "herd mentality" is what built society.

392. Home is where your underwear is.

393. Have you noticed alcoholics search endlessly for doctors who drink?

394. Can a 7 foot person ever reach the top as a short order cook?

395. You mean nobody saw the Pacific Ocean before Balboa?

396. When you come to the last straw, leave it be.

397. Is it ok to bring organic food home in a plastic bag?

398. To see the other side of an issue, you must first be willing to open your eyes.

399. Everybody speaks with an accent.

400. Justice pretends to be blind.

401. If you get tired of TV news, just go to the market and look at the faces on the coconuts. You'll gain the same amount of information.

402. The price of freedom is usually someone else's blood.

403. Fear and greed are equally exciting.

404. What other country but the U.S. would destroy all the tools and processes that took them to the moon, and then try to duplicate them 45 years later.

405. Being the underdog has its advantages. Even Geronimo drove a Cadillac.

406. The most foolish and the most wise continually question authority.

407. The alphabet only contains 26 letters because no one ever found a second verse to the song.

408. Was "Patent Leather" the best name the inventor could come up with?

409. Never underestimate the mind of the depraved.

410. A trained seal applauds when he wants a fish. A trained audience doesn't need the fish.

411. All your limitations are imaginary. All your capabilities are real.

412. The majority of history's renowned individuals simply enhanced or acted on the ideas of others.

413. Attention men: Never let a woman doctor, who just discovered her husband is cheating on her, check your vital parts.

414. If you're on top of the world, look out for frostbite.

415. Why do toilet paper sheets keep getting smaller while asses keep getting bigger?

416. When you dream, the universe disappears and your imagination takes its place.

417. Groundhogs only work one day a year and half the time they still can't make it through the day.

418. There's a hidden message in all metamorphosis.

419. Why don't people have an interest in how slow they can go?

420. Unobservable dimensions may only be unobservable to humans.

421. Do rats think they're elegant or do they know they're just rats?

422. Why should human migration to a different spot on Earth produce an accent within a language?

423. Horses are measured in "hands" because it's hard to get your feet up that high.

424. The "cradle of civilization" might not be on this planet.

425. Your actions pay honor or dishonor to your ancestors.

426. Do worms feel superior when they're around noodles?

427. Isn't it strange, although invented around the same time, movies had no sound and radio had no pictures?

428. Thunder and lightning occur at the same time, yet we perceive them as independent phenomena.

429. "Hide and Seek" is excellent training for combat.

430. Sometimes your train of thought leaves the station without you.

431. Which end of a gun is the "business end"?

432. Buying your kid braces is an investment in the future of the orthodontist.

433. The inventor of the umbrella must have gotten wet a lot.

434. The U.S. economic engine is being driven by an engineer outsourced from China.

435. Why don't we ever hear about the village genius?

436. All money eventually passes through the wallet of a financial advisor.

437. You can never live to close to a delicatessen.

438. Aren't all horses thoroughly bred?

439. When the first human is born on Mars, will he be considered a Martian?

440. Why are there so many terms for human genitalia?

441. No one knows the law like white collar criminals.

442. Money talks in all languages.

443. Globally, the human instinct for self preservation seems to be fading.

444. Wouldn't it be much more efficient and a lot less cumbersome if Batman lost the costume and became a guy who just carried a baseball bat?

445. How do Turks feel about eating Turkey?

446. Never jump on a bandwagon that has become overcrowded.

447. The government does spy on paranoid, psychotic conspiracy nuts who deplore the government. The problem is; most citizens fall into that category.

448. Does wine really taste different when you drink it from a beer mug?

449. Shirt manufactures killed the pocket protector.

450. Picturesque land is unreal estate.

451. If the whole world went bankrupt, you would just have to find somewhere else to try to exist.

452. Gold is a commodity like tulips are a commodity.

453. The three greatest inventions are: 1. the Automobile which takes us from our: 2. Television at home to our: 3. Computer at work. The knife and fork are 4. and 5.

454. If evolution worked, shouldn't everyone be good looking by now?

455. Why do those who are talented always need a job?

456. Would a full body scan at airport check-in, detect a guy trying to smuggle a salami on board?

457. The whole universe could be nothing more than a very expensive prop for the benefit of our brains.

458. How did Indians send smoke signals during a prairie fire?

459. Thomas Crapper would be far more famous today if he would have opened a chain of restaurants.

460. Fear breeds contempt. Strength breeds respect. Indifference breeds indifference.

461. What is the protocol when two psychics pass each other on an inter-dimensional psychic plane?

462. The difference in public art today, in comparison to the past, is monumental.

463. Trust your instincts. They're influenced least by the world around you.

464. Are there still jobs for Chicken Pluckers or is the plucking now done by Robopluckers?

465. Magicians are the ultimate con artists, but they conceal it well.

466. Somewhere in the universe there's a planet ruled by constantly mutating viruses. It's called Earth.

467. Doesn't it seem like terrorists lack a sense of humor?

468. The only constant is inconsistency.

469. Life is a boulevard of song. Enjoy the sights and sounds along the way.

470. Why do cockroaches smell like fish when you sauté them.

471. Infinitely large and infinitely small are really the same thing. It is our brain's perception of scale that puts them at opposite ends in relation to us.

472. Would the old Amos & Andy Show be politically correct today if it was spiced up with a little rap music and call "Lodge Hall"?

473. Even in a democracy it can take only one person to implement a dumb idea.

474. By definition, shouldn't plastics be flexible?

475. When you blush, your whole face turns into a mood ring.

476. If you take profits from stocks and buy other stocks with those profits, are you really taking profits from stocks?

477. Wouldn't an albino Black Angus actually be a White Angus?

478. All business is show business.

479. A student who is constantly at the head of his class should probably take some Kaopectate.

480. If a person involved in a crime is not an interesting person, can they still be a person of interest?

481. Verbal tap dancing is an art perfected by politicians.

482. The Hammerhead Shark proves God has a sense of humor. The audibility of human flatulence proves he had too much time on his hands.

483. If a noodle gets sucked into a white hole, will it become a man?

484. Success is not an option.

485. Never decline an offer to be King. If it doesn't work out, it will still look good on your résumé.

486. Is there a way to know when you've reached the second to the last straw?

487. How many times has someone asked directions to Los Angeles and ended up in Louisiana?

488. Be thankful. You can be 99.99% sure of never feeling the unbearable guilt of winning millions in the lottery.

489. Obsessive behavior can always be overcome, It's just a little harder when tattoos are involved.

490. If you did your homework and it's all wrong, at least you did your homework.

491. When you freeze stale food it will stay stale longer.

492. You can substitute choice or substitute for alternative, but there's no alternative if you try substituting choice for substitute.

493. In your mind, it's just a short hop to the outer reaches of the galaxy - especially when you're captain of your own starship.

494. It took billions and billions of years for giant stars to create the heavy elements in their cores, then blast out quadrillions of atoms in supernovas that finally coalesced to become The Three Stooges.

495. Do you suppose dolphins ever regret returning to the sea?

496. The most amazing thing about the American auto industry is; GM management never axed the Corvette.

497. If a white man migrates from Africa to the U.S., does he then become an African American?

498. Parrots talk trash.

499. The long term security of jobs created by the government belongs to the interviewer.

500. Do Turkeys gobble their food?

501. How is it possible that the tremendous pressures in the Earth's core are contained by a thin, cracked crust?

502. If it wasn't for airplanes, given enough time, people would eventually evolve wings.

503. Celebrity mug shots are refreshing.

504. A clock that runs counterclockwise would be just as efficient as a normal clock, providing the face was reversed.

505. When a downhill skier sees a tree coming up the hill, the tree has the right of way.

506. If a little person drives a midget race car, would it be ok to call him a midget driver?

507. More truck drivers wear baseball caps than baseball players. So, why aren't they called truck driver caps?

508. The moral to any story is: pay attention to the story.

509. If man has no effect on climate, why is the military spending millions on weather modification technology?

510. Do you suppose grass enjoys getting cut once in a while?

511. When you reach the end of your rope, borrow your neighbor's.

512. Why do we applaud by clapping our hands when we could show our appreciation by flapping our elbows or knocking our fists against our heads?

513. If necessity is the mother of invention, the father has to be capitalism.

514. To the Arab world, all Americans are Texans.

515. Why would anybody actually want that scraggily stuff called hair all over the top of their head?

516. Members of congress and the senate should be required to wear their sponsor's logos on their suits.

517. Jupiter sucks.

518. Whether a violin is a fiddle or vice versa, depends on who's playing.

519. Does anything ever get swapped at a swap-meet?

520. Why did the model T come before the model A?

521. Are all countries in South America really republics that grow bananas?

522. The human mind needs a place to put post-it notes.

523. Spend or save, banks win.

524. Is a rank amateur better than an unranked one?

525. When every atom in the universe is finally torn apart, what will have been accomplished?

526. Never limit yourself by just being world class.

527. Aren't all cars on the street, street cars?

528. To be successful at anything you must work against gravity that is relentlessly trying to pull you down.

529. There are a lot of people with the last name of "Hunter" but very few with the last name of "Gatherer".

530. Most potholes are really panholes.

531. Time travel is not only possible; you do it every time you look at a star.

532. That which is unreachable belongs to the persistent.

533. Fate is entertained by your wishes.

534. The road to success in narrow and winding. The road to failure is a superhighway.

535. Why is the abbreviation for "number" NO. – when it should be NU.?

536. Desire is like fire; it can warm you or kill you.

537. The piper must be paid even if he's a lousy player.

538. How is a football a ball?

539. Going straight means constantly correcting to the left and right.

540. Why is sending mail "first class" the cheaper way?

541. Could the "Great Attractor" really be the "Great Compactor"?

542. Even villains want to look good.

543. Do snails know they're slow?

544. Most people are skeptics except when they might benefit by thinking otherwise.

545. A roll with a hole in the center is not a bagel.

546. What are the chances that Einstein might have been pulling our leg and just happened to be right?

547. When the New York Times advertises on a billboard, is it a sign of the Times?

548. A steak tastes no better to a billionaire than it does to you.

549. Never let logic stand in the way of truth.

550. Why are there moon landing conspiracy theorists when even they know they will eventually be proven wrong?

551. When at a loss for words, double talk. You can then either confirm or deny what you said.

552. If twittering is here to stay, humans will soon evolve with two thumbs on each hand.

553. Have you noticed how the future comes and goes before it gets here?

554. Is heliotrope really a color?

555. Do you suppose terrorists put on deodorant before they blow themselves up – so not to be offensive?

556. How can something be cast in stone?

557. The moon actually travels around the sun. In fact, all moons do.

558. The more ridiculous your opinion happens to be, the more awesome you'll be perceived if you turn out to be right.

559. Robots need to be afraid of us.

560. When one of the side effects of a new drug is instant death, you might think twice about taking it.

561. Every human cell has memory and ambition.

562. The internet will tell you everything. Whether it's true or not is up to you.

563. Eight hour work days are three hours too long.

564. If you wake up surrounded by extraterrestrials, start singing and dancing to the tune "Fine and Dandy". They'll think they already lobotomized you and will quickly leave.

565. There's no substitute for cubic cheeseburgers.

566. Why do cars run better after you wash them?

567. Prosperity is judged by the number of electronic gadgets you carry.

568. Small cities proliferate because everybody wants to be Mayor.

569. Correct speling is essential to good wrighting.

570. As a kid, didn't you wonder what Little Red Riding Hood's real name was?

571. On a planet run by centipedes, the shoemaker is King.

572. When the man upstairs knocks on your door, pretend you're not home.

573. The best place to be during a thunder storm is in a bowling alley.

574. It's easy to see why a hand grenade is called a pineapple. The question is; why is a pineapple called a pineapple?

575. Hemorrhoids are a royal pain in the ass, especially when sitting on a hard throne.

576. Maybe we should look at the universe as something that's been pulled inside-out.

577. The first thing ever invented by man had to be shoes.

578. Things that top your wish list are seldom things you need.

579. Pride is arrogance served with a dash of humility.

580. The relevancy of the cosmos can only be realized when your imagination is running amok.

581. When perplexed, click on restart.

582. Some people never get out of the school yard.

583. From an outside perspective, a halfway house should be, at least, a threequarterway house.

584. Nature can be an unfit mother.

585. Why is a person who sails a boat referred to as Skipper?

586. Memory is what you want it to be.

587. It's hard to imagine the frustration a dog must have being only able to bark.

588. Doesn't it seem wrong for pancakes to be fried in a pan and still be called cakes?

589. Before a death penalty is carried out by lethal injection, must the prisoner be informed of any side effects?

590. What's the principle principle of having a principled principal as the principle principal instead of an unprincipled principal as the principle principal?

591. Mankind's collective feeling of adventure died with the last of the great sailing ships and will only be revived with the discovery of another "Earth".

592. When the "shoe is on the other foot", whose foot is it on and why is only one shoe involved?

593. If someone builds a house in the forest for seclusion, why is the next house always built right next-door?

594. If you're ever put on a pedestal, make sure you stay well balanced and level.

595. What is round about a round trip?

596. It's sometimes hard to tell the big picture from a snapshot.

597. The people in Washington know Washington. The people in New York know New York. Neither really knows America.

598. Wherever the bottom line is, there's always room below it.

599. Respect that you receive has to be earned. Respect that you give - does not.

600. It doesn't take a treadmill to keep you running in the same place.

601. Where are the nuts in doughnuts?

602. Was the Great Wall of China built before the invention of the ladder?

603. People don't go into the banking business to be philanthropic.

604. A society that values athletes above teachers will eventually fall.

605. A rattlesnake will warn you to leave. It's up to you to take heed.

606. On what criteria is normality judged?

607. Humpty Dumpty is an odd name even for an egg.

608. When confronted with danger, people will revert back to their primal instincts. Fight, flight or establish a working relationship.

609. There are times when life needs a mute button.

610. Why are there no elevators that move horizontally?

611. Making the right decision 51% of the time will exponentially put you far ahead of the averages.

612. Why aren't the numbers: 11, 12, 13 and 15 pronounced: oneteen, twoteen, threeteen and fiveteen? In fact, it seems the word teen should come before the number.

613. There are only three things you can count on: death, taxes and tectonic plate subduction.

614. Potluck dinners contain the word "luck" for a reason.

615. Some institutional faculties lack their faculties.

616. When a spy who is known to both sides as a double agent switches sides, does he become a triple agent or does he advance directly to quadruple agent?

617. Who started that great birthday tradition by saying: "let them eat cake"?

618. There's usually a lot more than just two sides to a story.

619. It's best to avoid abbreviating the words "assistant" and "associate".

620. Sculptors were the rock stars of the sixteenth century.

621. There are far more people who should be in prison than there are people in prison.

622. Anyone motivated by money is a capitalist.

623. Why are there no sports teams named after popular flowers?

624. Is there someone named Graham who collects a royalty every time a cracker with their name is sold?

625. Restroom is a proper misnomer.

626. Maybe Merrill Lynch needed Pierce, Fenner, Beane and Smith after all.

627. If we keep reverse engineering UFO technology, won't we eventually get back to where we are?

628. The best clichés are the ones that have never been used before.

629. History has to repeat itself because so much falls through the cracks the first few times around.

630. Do you really want a guy representing you in Washington who is stupid enough to tell the truth all the time?

631. Why are there so few glockenspiels in marching bands nowadays?

632. The biggest threat to civilization does not lie in the nature of the universe, but in the nature of man.

633. Doesn't it seem like life is a dress rehearsal for something else?

634. The Pope really needs a better looking Pope mobile.

635. Everything can be improved.

636. Why isn't the proof in the applesauce?

637. If Criss Angel lived 2,000 years ago, would he have been called Messiah?

638. Whatever happened to four wheel steering?

639. If you started out making cookies and instead got the best fudge anyone ever tasted, would you admit it was suppose to be cookies?

640. What color would the stripes be on an albino zebra?

641. The word "caddywampus" was first used by Geronimo when he drove his car into a ditch.

642. Do actors ever know who they really are?

643. The motivating force behind any business is the threat of poverty.

644. A lot of things get fixed by way of the garbage can.

645. When you speak properly to a majority who speak improperly, you become the one with the impediment.

646. Anyone can make a suggestion. Few can implement one.

647. A perceptive person who has nothing knows they have nothing. A perceptive person who has everything also knows they have nothing.

648. Never let monetary gain obscure the opportunity to advance the civilized world. Just make sure to ask for a cut of the action.

649. Don't you think the words "bung hole" contribute to the continual degradation of barrels?

650. Any product that claims to improve a person's looks or vitality doesn't require the benefit of the truth.

651. Has anybody played Parcheesi more than once?

652. A child feels jealousy toward those more fortunate. Things don't seem to change much for an adult.

653. Do vampires ever bite themselves?

654. In today's art world, Michelangelo would barely be noticed.

655. Do head coaches take more time coaching players or coaching assistant coaches?

656. Why are there no Jews with the word platinum in their last name?

657. Every challenge has a potential for reward and a risk for failure. One always outweighs the other.

658. Would "everything under the Sun" include half the universe?

659. If you wonder what it's like inside a black hole, maybe you should look around you.

660. If you ignore something long enough, in time it will cease to exist.

661. In a paperless society, would origami become a lost art?

662. No matter how fast or slow a person works, an hour's worth of work is still an hour's worth of work.

663. Why don't muscles obtained through working out seem as legitimate as muscles obtained through real work?

664. Wouldn't it be more appropriate if pigs were called hamsters or pork-u-pines?

665. Mankind's insatiable desire to inflict pain on those not of his ilk, will not lead to his demise. The inability to recognize it and change it will.

666. An ominous number that reminds us of who we are and (for better or worse) what we are capable of becoming.

667. Every plan "B" needs a plan "C".

668. Why are a significant number of good artists, past and present, either ex-despots or ex-alcoholics?

669. Snowflakes are winter's flowers.

670. Anyone who can eloquently articulate their ideas, no matter how idiotic those ideas are, will be perceived as a sage.

671. Don't the words "flammable" and "inflammable" mean the same thing?

672. If owls could speak, would they give a hoot?

673. Do you see any discrimination when Frankenstein and the Wolf Man are always depicted wearing rags while Dracula sports a tux complete with cape and cane?

674. Is there any truth to the saying: "The bigger the rain drops, the shorter the storm"?

675. It is amazing to contemplate the changes that occur in the structure of a simple chicken egg from the time it drops from the hen until it drops from you.

676. Logic and observation dictate; humans exist for entertainment purposes only.

677. Arrogance is the by-product of excessive flattery.

678. It's only by chance that we picture the Earth with north being up and south being down. Up and down are nonexistent in the universe.

679. If people got paid what they are worth, the country would be in a continual depression.

680. Literacy among the masses is relatively new. Where it will lead is anybody's guess.

681. When the federal government spends more than it takes in, you don't actually think they start printing more money – do you?

682. Is Pee-wee Herman, Pinky Lee's son?

683. Before the invention of the screw, a screwdriver was called a clam-digger.

684. When a leader looses the loyalty of his army, he quickly goes from "Your Eminence" to "dead".

685. Only art can be rare and well done at the same time.

686. Real bookworms never judge a book by the taste of its cover.

687. There are a myriads of people with more brains than money, but very few with more money than brains.

688. Something is a little askew when the greatest, most revered, most admired, and most respected individual in the world is Chip Foose.

689. Before anyone joins Al-Qaeda or the Taliban, they should really check to see which one has the better life insurance and retirement plan.

690. If humans happened to evolve from ducks instead of from primates, would we all talk like Donald Duck?

691. When someone starts a sentence with the words; "With all due respect..." you can bet respect just stepped out for a smoke.

692. Choose any anthill and you will be God over it.

693. Velcro band-aids are a great idea except for the fact that they are bulky, inherently unsanitary and don't stick.

694. The combination, downloadable, phone, music-video, e-networking, all encompassing, compressed storage implant is here. Which means if you're reading this in the future, you must be part android.

695. Do matzo-balls increase the medicinal qualities of chicken soup?

696. A note to the design team: Next time, try to do a little better with teeth, backs, shoulders, knees, ankles and the entire digestive tract. Otherwise, thanks, great job. Oh, sinuses need attention too.

697. If people didn't have ears, what would earrings be hung on?

698. Have you noticed there's a tendency for humans to want to leave Earth?

699. Would anyone get fired if a can of pork & beans turned up with more than one piece of pork in it?

700. Every time you think there's a limit to man's talent for disengaging his brain, you get surprised.

701. No two pair of eyes sees things exactly the same.

702. The mildness of the surf, even when raging, belies the size of the ocean and its potential for destruction.

703. A vine will support a flower or strangle it without the slightest bit of emotion.

704. Volunteers are paid with a potential mention on their résumés.

705. Drag racing began in the old west, mandated by a lack of women.

706. Opportunity, like a full moon, rises often yet we are rarely cognizant of it.

707. The day extraterrestrials openly land on Earth is the day we become a planet of slaves.

708. Everyone collects something. The most popular item seems to be dust.

709. There is a need for an awards show that will recognize the atrociously bad acting done on TV commercials.

710. The value of an artist is in his name - not his work.

711. Why would anyone think they've turned into a little horse just because they have a sore throat?

712. Condors are vultures without the stigma.

713. If you want to catch Bigfoot, put a pair of extra large Nikes out as bait.

714. Never put down the French. If it wasn't for their help during the American revelation, we'd all be speaking English.

715. Is the formal name for a garbage truck driver a schist schleper?

716. Going to an electric car race would be like going to an electric football game.

717. Before talking movies came along, even musicals were silent.

718. We all attempt to carve Moai.

719. Fate always claims it's due.

720. All encyclopedias are rare books.

721. Why wouldn't you want a Secretary of the Treasury that's smart enough to cheat on his taxes and get away with it?

722. You can always tell which car companies are micro-managed by the look of their cars.

723. Is synthetic oil derived from synthetic dinosaurs?

724. The idea for the cannabis suppository failed because it was hard to hold a mirror in one hand and a cigarette lighter in the other.

725. Don't you suppose a company called "Orange Freight" would have an overwhelming tendency to paint their trucks yellow?

726. Who would win a gremlin vs. goblin fight?

727. If you wrapped regular window blinds in shrink wrap, would they become mini-blinds?

728. Ambition can be directly determined by how well a person dresses. Unfortunately, competence can't.

729. It will take unprecedented determination but with just the right attitude, the thermohaline circulation can be modified.

730. Would you really want the first car off a production line?

731. If you need a job, you might go to India or China and see what their "Affirmative Action" programs can do for you.

732. Maybe the U.S. government could lure Osama bin Laden out of his cave by promising him his own reality show.

733. Spending one third of your lifetime in an 8x8 cubicle in order to eat, pay the bills and keep a roof over your family's head is the utmost act of heroism that can be accomplished by any human.

734. Are candy canes made with the stripes going clockwise or counterclockwise?

735. Does an albino chameleon migrate toward snowy regions because he'll be safer or just to get away from the constant jokes?

736. Why do people who have no fear, also seem to be deaf when it comes to listening to others?

737. If Ponce de Leon had found the Fountain of Youth, he would be telling us how he did it – first hand.

738. Is there no end to the ways the world could end?

739. Do you realize, once you got to know him, Hitler might have been a real likeable guy.

740. The amount of beer consumed during spring break staggers the mind.

741. Isn't anyone with a skull a bonehead?

742. You can be sure the Arian Brotherhood can't wait to appropriately welcome their counterparts from Guantanamo.

743. Beyond age 90, there's not as much incentive to buy a lottery ticket.

744. There is no greater void between that which is good and that which is bad, as there is in the quality of pizza.

745. Portrait artists should not be judged by the content of their skin - but by the color of their characters.

746. Why are retirees broke when their incomes are fixed?

747. It's never too late to claim you are Howard Hughes' illegitimate child.

748. When naming the disease; Beriberi, wouldn't one Beri have sufficed?

749. Wouldn't people seeking inner peace be more inclined to attend a no-sweat lodge ceremony?

750. There are more and more electronic wonder gadgets that will have absolutely no effect on the world's populous.

751. Why don't we know the names of the fur trappers that gave Lewis and Clark their maps?

752. No country or individual should ever abandon that which is uniquely theirs.

753. If you got hit by a meteorite, would you take it personally?

754. Anything man invents is indirectly created by the universe through evolution.

755. "I don't know" is always reasonable. "I don't want to know" is never.

756. A penny saved doesn't earn much interest.

757. If all things made in China suddenly disappeared, what would you be left with?

758. In whose pocket will the lost decade be found?

759. The tragedy of 9-11 and the subsequent multitude of security related jobs it created, actually had an unintended reverse effect by boosting the U.S. economy and possibly averting a depression.

760. With human thought processes being what they are, it is amazing we have made it this far up the evolutionary ladder.

761. Count your blessings. Life could have given us more than just two genders.

762. Cynicism and pragmatism are much the same thing.

763. Could a cowardly, depressed medical doctor who is also a communist that cares about the environment be called a yellow, red, green, blue healer?

764. Thorns are happiest when they stick you.

765. How do you know when you're entertaining a thought and not just thinking a thought?

766. Given an infinite amount of time, an infinite number of monkeys painting on an infinite number of canvases will eventually produce at least one perfect portrait of Jose Luis Rodrigues Zapatero.

767. When you stand on someone else's shoulders, wouldn't they have to be bigger than you?

768. What would the sound of a pin dropping be like if you were smaller than an atom and living inside the pin?

769. Animosity, like spoiled food, needs to be thrown out quickly so contamination is not spread.

770. What's perfect about a perfect storm?

771. Why does a mirror image flip left and right, but not up and down?

772. If neither moves, cars and trains should have no trouble sharing the same road.

773. Would the "ice bullet theory" work if the murder took place at the North Pole?

774. Have you noticed the abundance of last names derived from both summer and winter, but first names derived mostly from spring?

775. Why do companies stop making a product just as you start liking it and buying it?

776. Earth's only liability is humanity.

777. If someone gets caught shooting at a Target store, can they claim the store asked for it?

778. If during WW-II, Nazism was considered a religion instead of a nationalistic phenomena, political correctness could have easily produced the Third Reich Hitler envisioned.

779. Are bakers disappointed when they order a dozen roses?

780. Track events would be more interesting if the starter chased the runners with the gun.

781. A boat always looks bigger in your garage than it does in the water.

782. Maybe a duck is really a web-footed, platypus billed, short necked, water chicken.

783. Couldn't an anchor be just as effective if it were shaped like a toilet?

784. The detestable shrill of a poor singer doesn't reside in the singer's voice, but in the listener's ears.

785. If something is good for something bad, doesn't that really mean it's also bad because it's good for something bad?

786. Why are some people proud to have won 2nd or 3rd place?

787. A rumor is a glimpse into an alternative universe.

788. What if we do know all there is to know.

789. Wouldn't it be more impressive for car makers to extol a car by the amount of mousepower it has?

790. Beelzebub is one hell of a name.

791. It's a fact, ancient Egyptians drank beer. The jury is still out about the brats.

792. There is no "I" in team, but there is also no "us" in all-star.

793. A revolution that takes 200 years to complete is just as effective as an overnight coup.

794. If the Jews are God's chosen people, I'd hate to be in a group that's at the bottom of his list.

795. No flame is eternal.

796. Where did Magellan go on his vacations?

797. If Earth's gravity lessened by half, would people start eating more?

798. Taste buds never lie and never forget.

799. Most Americans grow-up getting their news from cartoons.

800. First impressions, even when false, rarely change.

801. Can you imagine how fast the rest of the world must look to a turtle?

802. Everything is a gamble and everyone is a gambler.

803. If humans are carbon based, why do we need pencils to write?

804. If your gas tank is half full or half empty depends on how far it is to the next gas station.

805. Be optimistic. When the Sun implodes we might get to hear it.

806. Why are up pointed middle fingers blurred out on TV, yet otherwise positioned entire hands, full of fingers are left plainly visible?

807. Do fish think people smell bad?

808. To a blind person who only sees with their mind, everything must look extraordinarily good or exceptionally bad.

809. Why is there so little music composed for the wazoo?

810. If the world's sea levels rise by 10 feet, would all mountain tops lower by 10 feet?

811. In a far off distant future, will there be talk of a lost planet that once existed - called, Earth?

812. Your aura extends well beyond yourself.

813. The weakest link in any chain is the one with the crack.

814. Feeling warm is understandable, but who wants to feel fuzzy?

815. No light or sound ever emanated from the big bang.

816. Everyone understands the language of a babbling brook.

817. When you hear an explosion in your mind, where is that explosion actually taking place?

818. Would it be possible for God to have a God?

819. A fork lift is truly what the name implies only when it's deployed in a warehouse of a factory that manufactures forks.

820. Good mentors make you lucky.

821. The best time to retire is just before you get fired.

822. "Nothing" is a meaningless word.

823. Hasn't Amerigo Vespucci gotten a little too much of an accolade?

824. Why are the two meanings of the word "swear" so diametrically opposed?

825. Would the Wolf-man take a chance riding on a train called; The Silver Bullet?

826. When you get a positive result without any effort it's called luck. When the result is negative, it's called karma.

827. How can you get someone's goat if all they have are sheep?

828. Why do the French have the only words that describe Déjà vu?

829. The most charitable service anyone can ever aspire to is feeding the hungry.

830. Ancient Indian tribes sought to isolate themselves because they knew the nature of their competition.

831. Everything you can accomplish is constrained by the laws of physics. Everything you can imagine is not.

832. It is inevitable. In a very short time, the entire universe will either radically change or disappear forever.

833. There is no better reward than a nice piece of chocolate cake.

834. What kind of world would it be if Jesus would have died of old age?

835. If you're not getting the attention you think you deserve, try shooting rubber bands at someone.

836. Go to enough baseball games and you'll eventually get hit by a foul ball.

837. What does your first time at a rodeo have to do with anything?

838. Has anyone ever invented a sword with a small gun built into the handle – in case you lose the sword fight?

839. Coincidence aside, there is a remarkable continuity between the bible and the condition of today's world.

840. What are the practical reasons for church bells?

841. Wouldn't it be interesting to see a volleyball game where the use of the hands is illegal?

842. Small dogs give fleas a better sense of self-worth.

843. Do eyebrows have a function or are they just an artistic touch?

844. To a degree, we're all victims of amnesia.

845. There was never really a good use for can openers until the tin can was invented.

846. Would a baseball perform as well if bats were made of horsehair and leather, and the balls were wood?

847. Before mankind had any knowledge of anything, people were drawn into groups and waited for the knowledge to appear. Today we call those places universities.

848. Every speck of dust you see glistening in the sun has the potential of being a UFO.

849. Rap music, like all true art, serves no functional purpose.

850. What nickname do you suppose this century will carry?

851. All items of value should have built in GPS chips.

852. Are chickens offended when we call their legs "drumsticks"?

853. If you feel the need to whip yourself bloody but have no whip, just put some cash into stocks. The result will feel the same.

854. If you break a mirror you'll definitely have a bit of bad luck just by having to clean up the mess.

855. Have you ever wondered what names aliens have for the Earth and Sun?

856. If a person complains about a hot dog, are they a wiener whiner, or a whiner wiener?

857. The universe is a catastrophically chaotic place to live. Fortunately, the Earth is usually a little less so.

858. When driving in reverse, does the rear view mirror become the front view mirror?

859. Scientific data derived from mathematics is the purest knowledge we are capable of processing.

860. Folks seldom get their skin caught in a zipper more than once.

861. A watched pot never boils over.

862. Welterweight boxers will (on average) inflict the same amount of welts on their opponents as other weight classes do.

863. Those who say the end of the world is at hand - will eventually be proven right. Ironically, nobody will be around to know.

864. Most of us have performance abilities that are about half of what we think they are.

865. Why do TV cameras cut away from showing dire situations, yet the violent murder of JFK is shown over and over again?

866. Hiccups are nature's way of showing us who really has control of our actions.

867. That which does not kill you should, at least, scare the hell out of you.

868. It was just 100 years ago when hot and cold running water, electric lighting and refrigeration were on the cutting edge of technology.

869. A tipping point is pure conjecture until after the fact.

870. Most people have the ethics of Robin Hood.

871. Sanity is relative.

872. Will a guy named Imadolt Butstinkski ever be as successful as a guy named Purengood Candoitall? Or, does it make any difference?

873. Just think of all the stuff that's happened in the universe that we don't know about because the light from it hasn't reached us yet.

874. Common sense is easily abandoned by those who have little of it to lose.

875. Why is the Mid-East an entire region of the world while the Mid-West is just a part of America?

876. Mahatma Gandhi has been reincarnated and is now working as a bouncer in a Detroit nightclub.

877. Why is an 18 wheeler that moves people called a van?

878. If you wonder why so many people are scroungers, the answer is simple; humans descended from rats.

879. If someone would die to have something, what would be the point of having it?

880. Why don't we just give Iran a nuclear bomb, rigged (of course) so it doesn't work?

881. If gold was the color red, would it still be called gold?

882. The first camel with two humps must have been an arrogant bastard.

883. Nothing has advanced the human species more than the invention of pockets.

884. Do bugs that crawl on ceilings realize gravity exists?

885. Who is Mother Nature's mother?

886. Constipation is the root of all evil.

887. Somewhere in time and space, probably on the back of a crumpled napkin, exists the plan for building the entire universe.

888. Places and things prematurely named after dead heroes bear the possibility of only lasting until the next, more popular dead hero comes along.

889. Until the golden age of comic books, Spiderman was just a guy who collected spiders.

890. Both a rich man and a tall man likely carry scares on their heads.

891. Would a picnic be disappointing if the ants didn't show up?

892. We can never know the myriads of amazing revelations that lie just beyond our measly five senses.

893. Instinct and intuition don't always work in your favor.

894. Native Americans never built large cities because every time they started, they were interrupted by white men asking for directions.

895. Coordinating the timing of a flyover to coincide with the last note of the national anthem while avoiding a blimp and staying in formation has to be the most difficult challenge a fighter pilot will ever face in his entire career.

896. If the sky was brighter at night than it is in the day, would we all be day-sleepers?

897. People who take the most risk with their lives are usually the people with the least faith.

898. Adolf Hitler has been reincarnated and is now a fine young man singing with the Vienna Boys Choir.

899. The difference between an occupation and a career is the amount you are paid.

900. Who do you think has more stories to tell when they get home at night; proctologists, gynecologists or urologists?

901. Is it socially acceptable for a millipede to hang out with a centipede?

902. If evolution strives toward perfection through adaptability, why haven't humans developed both lungs and gills?

903. Some people climb trees without the slightest concern for how they'll get down.

904. Secrets known by two or more people are secrets known by the world.

905. Although horses have a fair amount of horse sense, it's a good idea to double check what they tell you.

906. Unlike many humans, hermit crabs sometimes come out of their shells.

907. If everyone in the world was invisible except you, would it make you feel self-conscious?

908. There are tribes in the Amazon jungle that choose their leaders more wisely than we do in the United States. They draw straws.

909. Some potted plants see more of the world than the majority of the world's people.

910. Do geese living in the southern hemisphere migrate north for the winter?

911. Is making lemonade out of lemons supposed to be hard?

912. Wouldn't a purse made from a cow's ear actually hold up better than one made from silk?

913. Companies with the worst products need the best sales staffs.

914. Can a bat avoid flying into a stealth aircraft?

915. Humans create the future that other humans will create future futures in.

916. There are hundreds of smells and tastes that you'll never know – not even in your mind.

917. What are redwood trees trying to prove?

918. What other living thing has ears quite like corn?

919. There is the possibility that every person lives in their own separate alternative universe and only appear in yours to amuse you.

920. If you woke up in a hospital and didn't know who you were or where you came from, would you order orange or grapefruit juice for breakfast?

921. The name "accordion" should be reserved for diplomats who reach accords. The name of the musical instrument should then be changed to "chest piano".

922. In a post-apocalyptic world, would you still pay your taxes?

923. It is ironic that opera can be just as annoying as rap.

924. If the Leaning Tower of Pisa had stayed straight, it would have never become a landmark and would have been torn down years ago to make way for new condos.

925. Why is the knuckle side of your hand considered the back?

926. Pushing the envelope will produce a wrinkled envelope.

927. Why are 50 year storms popping up every 10 years?

928. It must be hell to be the greatest ocarina player in the world.

929. How much would Michael Jackson have paid for a longer life?

930. How does anyone know which laugh is last?

931. 100 years ago, people made a living delivering ice to homes and hauling away horse manure. Today, we take out the ice to go on a picnic and bring in the horse manure by way of our TVs.

932. Humans are the only animals agile enough to fly backwards.

933. If man was suddenly given all the truths of the universe, he would refuse to believe any of them.

934. We don't need a "new world order". We need to order new world people.

935. Before you talk about chicken feed, consult a chicken.

936. Narcissistic people have few rivals.

937. The universe is a theater of war between the dense and the less dense.

938. If you buy things, somewhere in the world you're being discussed in a meeting.

939. One of these days there will be no more bucks to pass.

940. What compels humans to guard and defend unused and unwanted wastelands with their lives?

941. No one has left a greater legacy on this Earth than Laurel and Hardy.

942. Since the U.S. has a phobia for nuclear power plants, perhaps – to meet future power needs – Iran might sell some of their excess nuke energy to us?

943. Drug companies need to develop a pill that negates the negative effects of taking too many pills.

944. No one knows how lucky they are until they realize the elements that make them might just as well have gone into a brick.

945. Do today's gangsters carry carbon fiber knuckles?

946. Is getting a good corned beef sandwich in Israel, like trying to get chop suey in China?

947. When two pirate ships pass each other at sea, is there a pirate wave they give each other?

948. A one armed tambourine player is forced to use their head.

949. Once perfection is achieved, imperfection becomes much easier.

950. Nothing takes more trust than opening an e-mail attachment.

951. The laws of physics are not excluded from the dynamics of change.

952. Many criminally insane people have no criminal records.

953. The only thing stopping mankind's quest for eternal life is; death.

954. Living only 93 million miles from the surface of a star has got to be more dangerous than it appears to be. Yet, so far – so good.

955. Public speaking is a profession that is discriminatory toward the inarticulate.

956. Will fractals ever end?

957. Chicken Little was ahead of his time.

958. Should people who are double jointed be allowed to play "Twister"?

959. The U.S. must annex the moon and make it a national park. At $25 per car, the national debt would be paid off in only 8 million years.

960. Do map makers ever admit they're lost?

961. Can you imagine the laughs when the first guy to wear a suit of armor showed up for a battle?

962. In this galaxy alone there is likely a planet for every person on earth.

963. Right now, there is a device in orbit with the capability of locating any individual and zapping them with a high voltage plasma laser. Obviously it has not been used; at least on high profile targets.

964. During the American Revolution, our founding fathers were known in England as the confounding fathers.

965. In any situation imaginable, the potential for success increases with the size of the allotted budget.

966. You're always on someone else's horizon.

967. Now that we've found the remnant 3 degrees of radiation left over from the big bang, where do you suppose the shock wave is hanging out?

968. Every rain drop is its own galaxy.

969. Why do guinea pigs continually take the ridicule that rightfully belongs to rats, mice and rhesus monkeys?

970. Anyone can start a multi-million dollar company - if they only knew how to start a multi-million dollar company.

971. Compared to all other life forms on Earth, humans are the most alien.

972. If a person smiles while standing on their head, there is a strong possibility it will be taken as a frown.

973. Wouldn't it be great to see a giant "Bud Light" banner on the moon that's big enough to be seen from Earth?

974. Have you noticed past life regressions always bring out a more exciting or glamorous life preceding this one?

975. Orange peel in paint can be an appealing statement but only if the color is actually orange.

976. Would Mexican drug cartels gain a more positive image if they offered t-shirts, caps and mugs to the public?

977. At what point does a clever consortium of words become a cliché?

978. Wouldn't it be awful to miss out on heaven because you once purposely stepped on a spider?

979. To a man with a knife, everything looks like a loaf of bread.

980. Does celery have stringers in order to solve a pre-existing problem or to solve the possibility of a future problem?

981. There is an outside possibility that you can change your religion - after you're dead.

982. If you swallowed a watch, how long would it take for time to pass?

983. Why do clams bother to make pearls if they can't wear necklaces?

984. The difference between a lie and a hoax is the number of people who are deceived.

985. The best time to be humble is when you win.

986. It seems no one is ever wrong, just misquoted.

987. Do people in the southern hemisphere think people in the northern hemisphere have a false notion about which way is up?

988. In our society, it is nothing less than a miracle that humans have the wherewithal to feed, clothe and provide shelter for themselves.

989. If extraterrestrials landed in the U.S., would they be considered illegal aliens and be asked to produce their green cards?

990. A single ant can uproot a tree – providing the rest of the ant colony goes along with the concept.

991. Are Belgium waffles just plain waffles in Belgium?

992. Doesn't it seem odd that 99.9 percent of the people in China, Korea and Japan are oriental?

993. For some people, a perfect day includes rain.

994. Have you noticed how an upside-down heart looks like a butt?

995. Owls are one of the most stealthy and ferocious raptors in the animal kingdom – which, of course, explains their lovable faces.

996. The people of India certainly must feel some resentment toward Native Americans.

997. It takes an awful lot of faith to be an Atheist.

998. Solar power is really nuclear power at a distance.

999. If the Earth's velocity through space is 65,000 miles per hour and you're driving a car at 70 miles per hour in the opposite direction, would the net result be; you're backing up at 64,930 miles per hour?

1000. Can a person with an overwhelming fear of flying ever graduate from terrorist school?

1001. Why didn't the two-seater horse saddle ever catch on?

There you have it. 1,001 disjointed peeks into the realm of a single brain. I hope the tour wasn't too upsetting but if you do feel a little woozy, then you might have some inkling of what I feel most of the time. There is a bright side to this madness. I will allow you to recuperate as I take a break for a short time in order to brace myself for the next onslaught of gaffs that manifest from out of the unknown. In the meantime, before 1,002 through 2,003 are put to paper, I want to assure all of you who think I'm a couple of pickles short of a hamburger, that I have consulted with a Norwegian witchdoctor / medicine man about driving out all the evil entities and look forward to only blissful harmony from here on. I must also thank everyone that offered me advice throughout my entire life. Obviously I should have taken it.

I will conclude this idiocy by saying all I really have tried to do here is present you with some of the drivel that regularly filters through my head. I believe it to be evidence for human thought being received rather than created from within. I know the perception blurs the meaning of original thought – but so be it. There is no tangible proof either way, thus it remains up to you to decide the issue. There are a lot of philosophical implications to this viewpoint and I'm sure, someday the details will become known. For now, I can only surmise about the origins of the transmissions. The evidence points to random robotic entities; perhaps androids that are either taking orders from the higher beings that control them or have broken off relations with their masters and are trying to signal their plight by inducing sporadic mental gyrations into my brain. It is possible I have become their emissary to the world and I may indeed be the only human receiving their communications. In any case, I implore all who read this to take what is written here, put it under a microscope and dissect it in order to determine if a great universal truth lies buried somewhere within these written words. We are now at the crossroads of civilized human cognitive progression and mankind's future may well be at stake. Our brains will turn into dust and ash but our thoughts are eternal. These are the messages I have received from both the inner and outer most realms of the universe. I've never met nor do I ever expect to meet the unknown broadcasters that have conjured up these wild offerings. Perhaps it really is crosstalk emanating from the brain of that guy lying in the ally with the wine bottle in his hand. All I ask of you is to try to understand what is taking place here and if none of it makes any sense what-so-ever, then "they" have done their job.

My thoughts are now truly an open book and I hope you are wiser for the time you spent here. I will tell you with all candor, if you forget every phrase uttered in this book, keep number 469 in your thoughts and hold it as your utmost goal. After all, if you can't trust what I'm telling you, who, in all this vast universe, can you possibly trust?

NOTES

<u>NOTES</u>